Marco Aurélio Schiavo Novaes
Cristina Mattos Veloso

Creep-feeding lambs with diets containing castor bean meal

Marco Aurélio Schiavo Novaes
Cristina Mattos Veloso

Creep-feeding lambs with diets containing castor bean meal

Federal University of Viçosa, 2015

ScienciaScripts

Imprint

Cover image: www.ingimage.com

This book is a translation from the original published under ISBN 978-3-330-75620-5.

Publisher:
Sciencia Scripts
is a trademark of
Dodo Books Indian Ocean Ltd. and OmniScriptum S.R.L publishing group

120 High Road, East Finchley, London, N2 9ED, United Kingdom
Str. Armeneasca 28/1, office 1, Chisinau MD-2012, Republic of Moldova, Europe
Printed at: see last page
ISBN: 978-620-8-27629-4

SUMMARY

ACKNOWLEDGMENTS

I thank God, first and foremost, for entrusting me with this responsibility. For the countless opportunities to grow and learn. For all the help and affection dedicated to me, in moments of trial and abandonment, but also in moments of true happiness, which were always there.

I would like to thank my parents, my eternal friends and teachers, my companions in struggle and ideals. To you, who always did your best, who never settled for "good enough". Eternal gratitude for all the encouragement, for all the advice and care, for every minute of effort in consolidating the human being I am today. For all the love and dedication, I thank you from the bottom of my being.

I would like to thank all the teachers who have contributed in some way to my academic development. Without your work, this would never have been possible. My special thanks go to my advisor, Professor Cristina Mattos Veloso, for her trust and commitment to this project.

I would also like to thank the scholarship holders, interns and friends who contributed so significantly to the smooth running of the project, both in the field and in the laboratory. I would like to thank you and your willingness to help whenever possible.

Finally, I would like to thank the Federal University of Viçosa and FAPEMIG for their enormous contribution, granting us financial aid and providing us with the use of space, materials and equipment to carry out and analyze the project. Also to CNPq for the grant.

SUMMARY

NOVAES, Marco Aurélio Schiavo, M.Sc., Federal University of Viçosa, February 2015. **Supplementation *of creep-feeding* lambs with diets containing detoxified castor bean meal.** Supervisor: Cristina Mattos Veloso. Co-supervisors: Giovanni Ribeiro de Carvalho and Clàudio José Borela Espeschit.

Data suggesting levels of substitution of soybean meal by castor bean meal are scarce for feeding *creep-feeding* lambs on pasture. The aim of this study was to evaluate the replacement of soybean meal with detoxified castor bean meal, in terms of consumption and digestibility, weight gain, carcass yield, physiological and urinary parameters; the effectiveness of *creep-feeding;* and possible tissue damage resulting from the inclusion of castor bean meal. Forty-three male lambs were used, with an average initial weight of 7.95 kg, from 15 to 75 days of age, obtained by crossing Dorper rams with Santa Inês ewes. The statistical design used was a randomized block design, where the blockage was based on the date of entry into the experiment. Five treatments were used, sequentially: without supplementation*;* with supplementation, but without the addition of castor bean meal; and gradual levels of substitution of soybean meal with detoxified castor bean meal (33, 67 and 100%). The statistical procedures, contrast, mean and standard error of the mean, were carried out using the Mixed Procedure program (PROC MIXED), using the statistical package SAS (*9.3*), adopting $\alpha = 0.10$. Replacing soybean meal with detoxified castor bean meal, under the conditions of the experiment, did not significantly affect the performance of the animals, while the use of *creep-feeding* had a positive effect, with the gain of the supplemented animals being higher than the control. The intake of detoxified castor bean meal resulted in a linear increase in the number of eosinophils. Concentrate consumption has a negative influence on the thickness of the papillary epithelium of the ventral kidney sac.

GENERAL INTRODUCTION

Sheep farming has great potential in the national livestock sector, as there is a deficit in sheep meat production, since part of Brazil's demand is supplied by imported meat, mainly from Uruguay (Barreto Neto, 2007; Medeiros et al., 2009). Landin et al. (2007) state that the sheep meat production chain in the country is still in its infancy.

In 2012, there were 16.789 million head of sheep, with the Northeast region accounting for 55.5% (10.112 million head) of the national herd (IBGE). The national agro-industrial chain is not efficient, often offering animals slaughtered at an uneven weight and late, with low quality meat and less consumer acceptance (Macedo et al., 2008).

Producers have been increasingly looking for technologies to optimize production processes and supply the increasingly demanding market with high-quality meat (Osório et al., 2012).

This market profile, which demands younger animals with better quality meat, encourages the intensification of lamb finishing systems, but can lead to an increase in production costs. This premise supports the search for alternative feedstuffs, such as agro-industrial by-products, for animal nutrition, with the aim of reducing costs (Carvalho et al., 2012) and meeting an increasingly demanding market in terms of the sustainability of systems. It is important to consider that ecologically correct production techniques can also result in a differentiated product on the market.

The creation and adaptation of technologies that cater for this niche market, reducing production costs and improving productivity, are promising. Sheep confinement has been adopted as a method of intensifying the production system, reducing the mortality of offspring due to nutritional deficiencies and, in some cases, parasitic infestations, as well as the regularity of the meat supply throughout the year.

As part of the cost of producing lamb, concentrated feed has the biggest impact, accounting for 60 to 70% of total costs. Therefore, the search for alternative feeds that can replace traditional feeds, such as corn and soybean meal, has become increasingly frequent, with the aim of reducing production costs without causing a loss in performance, since traditional feeds are subject to market fluctuations (Parente et al., 2009; Santos, 2011).

One alternative is the use of co-products from biodiesel production in ruminant feed. However, the effects of these foods on nutrition at certain stages of sheep rearing still need to be better elucidated.

The production of biodiesel is in line with global concerns about the production of energy with a lower environmental impact, since it is a renewable and less polluting source of fuel

(Abreu et al., 2004). The increase in oilseed production for this purpose results in large quantities of solid waste, whether cake or bran. The integration of the agro-energy industry and livestock production, through the use of waste from oil extraction in animal production, promotes the optimization of jobs and income generation in these sectors, as well as reducing environmental liabilities, since a potential pollutant (cake or bran) is now used as an input for the production of animal feed (Oliveira et. al, 2010).

Among the various oilseed options available for biodiesel, the castor bean crop (*Ricinus communis L.)* is characterized by its potential for exploitation in regions marginalized from the development process and by its high potential oil productivity per unit area, when compared to annual cycle oilseeds (Barros et al., 2006; Ferreira et al., 2006; CONAB, 2008). This is due to its high phenotypic plasticity, which allows the selection of plants adapted to different soil and climatic conditions, making it possible to grow it almost anywhere in the country (Ferreira et al., 2006).

In 2014, castor bean production (in pods) amounted to 55,710 tons, a positive variation of 365.9% compared to the previous harvest. Of the total produced in the last harvest, 96.2% was produced in the Northeast, which expresses the great potential of castor beans in this region (IBGE, 2014). If all castor bean production had its oil extracted by solvent, castor bean meal production would have been 27,855 tons in the last harvest. This represents the great potential of this crop to provide protein in animal feed.

Castor bean bran is obtained using chemical solvents to extract the oil, which allows for a product with a lower oil content (below 1.5%) and, consequently, a higher protein content (Evangelista et al., 2004), characteristic of protein concentrate foods, which can correspond to 80% of the crude protein content of soybean meal (Moreira et al., 2003).

Castor bean bran should only be used after detoxification (Abdalla et al., 2008) of the potentially toxic compounds ricin (alkaloid), CB-1A (allergenic complex) and ricin (protein) (Severino, 2005), the latter being the only one capable of causing poisoning in animals when ingested.

In pasture systems, weight gain per animal and per area is strongly influenced by the daily availability of dry matter and the stocking capacity of the pasture (Carnevalli et al., 2001), as well as the quality of the forage and animal consumption.

In order to maximize production results in sheep farming, genetic and environmental improvements must be promoted. To this end, in addition to cross-breeding, zootechnical indices must be improved, especially by reducing the age at which lambs are slaughtered, with creep-feeding being an option in this context (Santra and Karim, 1999).

The use of *creep-feeding* is justified by the fact that the lactation curve of ewes is characterized, after the peak of lactation (third and fourth weeks of lactation), by a progressive reduction in milk production concomitant with an increase in the nutritional needs of the young. This imposes the need to correct nutritional deficits. Furthermore, when ewes have more than one lamb, the increase in milk production is not proportional to the increase in the number of young, thus generating a nutritional imbalance. Another factor that contributes to the imbalance of nutrients for the young is the decline in the quantity and quality of forage in the pastures throughout the year. In the dry months, pasture does not normally meet the requirements of lambs in the rearing phase (Viana et al., 2004).

Lambs should be *fed creep-feeding* from seven to ten days of age, although consumption is still low up to two or three weeks of age. Even if consumption is low, it is important to keep the *creep* feed fresh, whether supplemented with concentrate or roughage (Viana et al., 2004).

According to Silva Sobrinho (1997), good feeding during the suckling period increases the rate of growth, reduces mortality and avoids restrictions on the animal's future production. As an additional gain, the *creep-feeding* system in the rearing phase can provide a shorter lactation period, releasing the dams for breeding earlier, as well as being able to provide lambs for slaughter in less time than the traditional suckling system, including more uniform batches.

The use of detoxified castor bean bran in ruminant feed is already well established in the literature, when supplied to the rearing and finishing stages. However, there is little information on how to feed it to suckling lambs.

The aim of this study was to evaluate the replacement of soybean meal with detoxified castor bean meal in terms of consumption and digestibility, weight gain, carcass yield, physiological and urinary parameters; the effectiveness of *creep-feeding*; and possible tissue damage resulting from the inclusion of castor bean meal.

BIBLIOGRAPHICAL REFERENCES

ABDALLA, A.L.; SILVA FILHO, J.C.; GODOI, A.R. et al. Utilization of by-products from the biodiesel industry in ruminant feed. **Revista Brasileira de Zootecnia**, v.37, special supplement, p.260-258, 2008.

ABREU, F.R.; LIMA, D.G.; HAMÙ, E.H. et al. Utilization of metal complexes as catalysts in the transesterification of Brazilian vegetable oils with different alcohols. **Journal of Molecular Catalysis A: Chemical**, v.209, p.29-23, 2004.

ANANDAN, S.; ANIL KUMAR, G.K.; GHOSH, J. et al. Effect of different physical and chemical treatments on detoxification of ricin in castor cake. **Animal Feed Science and Technology**, v.120, p. 159-168, 2005.

BARRETO NETO, A.D. Strategic positioning of the goat and sheep meat sector in the Brazilian meat market. In: SIMPÓSIO INTERNACIONAL SOBRE CAPRINOS E OVINOS DE CORTE, 3., 2007. **Proceedings...** Joao Pessoa: EMEPA, 2007. CD-ROM.

BARROS, G. S. de C.; SILVA, A. P.; PONCHIO, L. A. et al. Biodiesel production costs in Brazil. **Revista de Politica Agricola**, Ano XV - n° 3 -Jul./Aug./Sept. 2006.

CARNEVALLI, R.A.; SILVA, S.C.; FAGUNDES, J.L. et al. Sheep performance and pasture responses to Tifton-85 (*Cynodon* spp.) under continuous stocking. **Scientia Agricola**, v.58, n.1, p.7-15, 2001.

CARVALHO, H.H.; JONG, E.V. **Alimentos: métodos fisicos e quimicos de análisis.** Porto Alegre: UFRGS, 2012. 180 p.

CONAB - National Supply Company. 2008, available at www.conab.gov.br.

EVANGELISTA, A.R.; ABREU, J.G.; PERON, A.J. Evaluation of the chemical composition of castor and peanut cake obtained by different oil extraction methods. In: Brazilian Castor Bean Congress, Campina Grande, 2004. **Proceedings...** Campina Grande: Brazilian Agricultural Research Corporation. Available at: www.cnpa.embrapa.br.

FERREIRA, G.B.; BELTRÂO, N.E.N.; SEVERINO, L.S. et al. **Castor bean cultivation in the cerrado: risks and opportunities.** Embrapa Algodao: Campina Grande, 70p, 2006 (Documentos, 149).

IBGE - BRAZILIAN INSTITUTE OF GEOGRAPHY AND STATISTICS. **Systematic survey of agricultural production**. Rio de Janeiro, v.27, n.1, p.185, January/2014.

. **Municipal Livestock Production.** Rio de Janeiro, v.40, p.1-71, 2012. ISSN - 0101-4234.

LANDIN, A.V.; MARIANTE, A.S.; McMANUS, C. et al. Quantitative carcass characteristics, morphometric measurements and their correlations in different sheep genotypes. **Ciência Animal Brasileira**, v.8, n.4, p.665-676, 2007.

MACEDO V.P.; GARCIA C.A.; SILVEIRA A.C. et al. Tissue composition and loin chemistry of lambs fed diets containing sunflower seed in private feedlots. **Revista Brasileira de Zootecnia**, v.37, n.10, p.1860-1868, 2008.

MEDEIROS, G.R.; CARVALHO, F.F.R.; BATISTA, A.M.V. et al. Effect of concentrate levels on carcass characteristics of Morada Nova sheep in confinement. **Revista Brasileira de Zootecnia**, v.38, n.4, p.718-727, 2007.

MORAES, J.C.F.; SOUZA, C.J.H. de; GONÇALVES, P.B.D. Control of oestrus and ovulation in cattle and sheep. In: **Biotechniques applied to animal reproduction**. Sao Paulo: Varela, 2002, c.3, p.25-56.

MOREIRA, J.F.C.; RODRIGUEZ, N.M.; FERNANDES; P.C.C. et al. Protein concentrates for cattle. 1. in situ digestibility of dry matter and crude protein.

Arquivo Brasileiro de Medicina Veterinària e Zootecnia, Belo Horizonte, v.55, n.3, jun., p.315-323, 2003.

OLIVEIRA, A.S.; CAMPOS, J.M.S.; OLIVEIRA, M.R.C. et al. Nutrient digestibility, nitrogen metabolism and hepatic function of sheep fed diets containing solvent or expeller castorseed meal treated with calcium hydroxide. **Animal Feed Science and Technology**, v.158, n.1, p.15-28, 2010.

OSÓRIO, J.C.S.; OSÓRIO, M.T.M.; GONZAGA, S.S. et al. Lamb finishing. **Pub Vet**, Londrina, v.6, n.23, ed.210, Art.1402, 2012.

SANTRA, A.; KARIM, S.A. Effect of protein levels in creep mixture on nutrient utilization andgrowth performance of pre-weaner lambs. **Small Ruminant Research**, v.33, p.131-136, 1999.

PARENTE, H.N.; MACHADO, T.M.M.; CARVALHO, F.C. et al. Productive performance of sheep in confinement fed different diets. **Arquivo Brasileiro de Medicina Veterinària e Zootecnia**, v.61, n.2, 2009.

SANTOS, T.R. **Treated castor bean bran, dry and wet, in lamb feed.** Dissertation (Master's Degree in Animal Science) - Viçosa: UFV, 2011. 39p.

SILVA SOBRINHO, A. G. **Criaçâo de ovinos.** Jaboticabal: FUNEP, 1997. 230 p.

VIANA, J.C; BILLON, N.; CUNHA, A.M. The thermomechanical environment and the mechanical properties of injection moldings. **Polymer Engineering and Science**, vol. 44, pp. 1522-1533, 2004.

SUPPLEMENTING *CREEP-FEEDING* LAMBS WITH DIETS CONTAINING DETOXIFIED CASTOR BEAN MEAL

Abstract - Castor bean meal, when detoxified, can be used in animal nutrition as a protein concentrate, with crude protein levels similar to those of soybean meal. Forty-three male lambs were used, with an average initial live weight of 7.95 kg, from 15 to 75 days of age, obtained by crossing Dorper rams with Santa Inês ewes. The lambs were distributed according to a randomized block design, where the block was the day of entry into the experiment. Five treatments were used, sequentially: without supplementation; with supplementation, but without the addition of castor bean meal; and gradual levels of substitution of soybean meal with castor bean meal (33, 67 and 100%). The statistical procedures, contrast, mean and standard error of the mean, were carried out using the Mixed Procedure program (PROC MIXED), using the statistical package SAS (*9.3*), adopting $\alpha = 0.10$. The aim of this study was to evaluate the replacement of soybean meal with castor bean meal, in terms of consumption and digestibility, weight gain, carcass yield, physiological and urinary parameters; the effectiveness of *creep-feeding*; and possible tissue damage resulting from the inclusion of castor bean meal. To this end, samples were taken of the feed and leftovers, as well as the animals' feces, urine and blood. At the end of the 60-day experiment, the lambs were subjected to a 16-hour water diet, then they were stunned and slaughtered to weigh the carcass, limbs, skin, head, blood and viscera. The small intestine and its components were also measured separately and histological fragments of the ileum and ventral sac of the rumen were collected and analyzed. It was concluded that replacing soybean meal with detoxified castor bean meal did not significantly affect the animals' performance, while the use of *creep-feeding* had a positive effect, with the gain of the supplemented animals being higher than the control. The intake of detoxified castor bean meal leads to a linear increase in the number of eosinophils. Concentrate consumption has a negative influence on the thickness of the papillary epithelium of the ventral kidney sac.

Keywords: performance, digestibility, sheep - feed and rations

INTRODUCTION

Among the various oilseeds used to produce biodiesel, castor bean stands out for its significant production capacity (Barros et al., 2006; Ferreira et al., 2006). The agrochemical industry uses two oil extraction methods: physical (pressing) and chemical (solvent). Both generate solid waste. The former produces castor bean cake and the latter castor bean meal (Evangelista et al., 2004).

When ingested, castor bean meal can cause poisoning due to the presence of the toxic component ricin, and the animal may suffer abdominal pain, vomiting, severe dehydration, decreased urine production and lower blood pressure (Doan, 2005) when intoxicated. Specific damaging effects on the kidneys and intestines, for example, are not reported in the literature.

Castor bean meal, when detoxified, can be used in animal nutrition as a protein concentrate, with crude protein levels similar to those of soybean meal, which is the most common option in ruminant feed formulation (Moreira et al., 2003; Abdalla et al., 2008).

The commercial use of castor bean bran in animal feed has been common for decades. However, research is focused on rearing and finishing, limiting the extrapolation of conclusions to more specific categories, such as rearing. Therefore, nutritional management techniques that cater for this category need to be tested. In general terms, maximizing performance in this phase results in intensification of the production system, allowing lambs to be weaned early, at 60 days or less, and can be used from the 10th day of life. In addition, the intake of solids is important to stimulate the development of the kidneys at this stage, and should be conducted in a controlled manner, through the use of *creep-feeding*, for example (Susin, 1996).

It is therefore believed that castor bean meal, fed in *creep-feeding*, can fully replace soybean meal as the main source of protein in the diet of lactating lambs, without any compromise to the animals' performance.

The aim is to evaluate the replacement of soybean meal with detoxified castor bean meal, in terms of consumption and digestibility, weight gain, carcass yield, physiological and urinary parameters; the effectiveness of *creep-feeding*; and possible tissue damage resulting from the inclusion of castor bean meal.

MATERIAL AND METHODS

2.1 - Ethics Committee

UNIVERSIDADE FEDERAL DE VIÇOSA
CENTRO DE CIÊNCIAS AGRÁRIAS
DEPARTAMENTO DE ZOOTECNIA

Campus Universitário - Viçosa, MG - 36570-000 - Telefone: (31) 3899.2262 - Fax: (31) 3899.2275 - E-mail: dzo@ufv.br

Comitê de Ética para Uso de Animais

Viçosa, 30 de Março de 2011

CERTIFICADO

O Comitê de Ética para Uso de Animais do Departamento de Zootecnia da Universidade Federal de Viçosa certifica que o **processo nº 03/2011**, intitulado "**SUPLEMENTAÇÃO DE CORDEIROS EM CREEP-FEEDING COM DIETAS CONTENDO FARELO DE MAMONA**", coordenado pela **Profa. Cristina Mattos Veloso**, está de acordo com os princípios éticos da experimentação animal, estabelecido pelo Colégio Brasileiro de Experimentação Animal e com a legislação vigente, tendo sido aprovado por este Comitê em **30/03/2011**.

Marcos Inácio Marcondes
Presidente do CEUA/DZO/UFV

2.2 - Ricin analysis

The presence of ricin in castor bean bran was assessed according to the procedure described by Oliveira et al. (2010). Ricin was extracted from the bran using 0.5 M Tris HCL buffer, pH 3.8 (Oliveira et al., 2010). To identify the ricin fractions, electrophoresis was carried out on a 15% acrylamide/bisacrylamide gel (29:1) under denaturing conditions (SDS-PAGE), according to the method proposed by Laemmli (1970). The gels were stained with silver nitrate solution (0.2 g/L) and another with Coomassie Brillant Blue G solution (1.5 g/L) (Figures 1 and 2). The ricin A (approximately 36 KDa) and B (approximately 29 KDa) fractions were identified using molecular mass markers between 14 and 66 KDa (Sigma, USA). The detoxification method was considered effective only when the two ricin fractions had completely disappeared (Oliveira et al., 2010).

Staining with coomassie blue was more evident than with silver nitrate, although both showed the same result. Treatment with calcium hydroxide proved to be effective in denaturing the ricin alkaloid, since, when compared to the standards, only the aliquots of untreated castor bean (MND) showed the compound, while in the two aliquots of the two treated castor bean samples (MNBa and MNMg) ricin was not present.

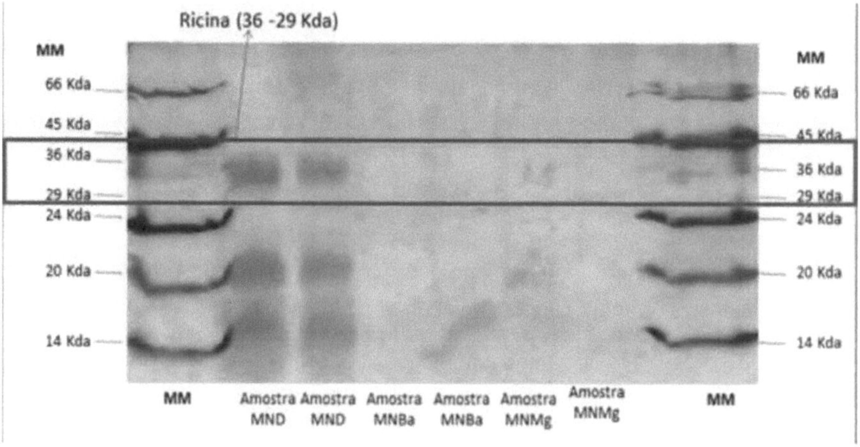

Figure 1. Gel stained with silver nitrate.

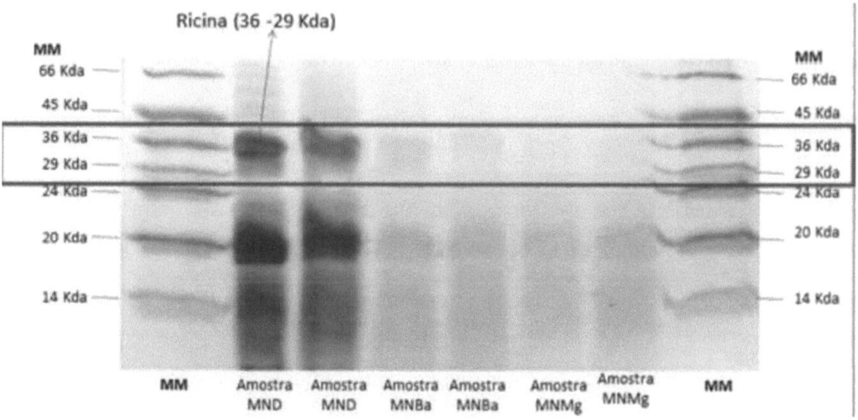

Figure 2: Gel stained with Coomassie Brillant Blue G.

1.3 - Animals, treatments and food

Forty-three male lambs were used, with an average initial live weight of 7.95 kg, from 15 to 75 days of age, obtained by crossing Dorper rams from a commercial farm and Santa Inês ewes from the UFV flock.

The detoxification of castor bean bran was carried out according to the methodology proposed by Anandan et al. (2005), where 100% of the ricin was inactivated with $Ca(OH)_2$ at a dose of 40 g/kg of bran or with an autoclave at 15 psi for 60 minutes. Oliveira et at. (2010), evaluating dosage and treatment, detoxified castor bean bran using alkaline treatment (calcium hydroxide - $Ca(OH)_2$, or calcium oxide - CaO, at doses of 20, 40 or 60 g/kg, diluted or not in water) plus resting the material for 18 hours, or thermal treatment (autoclaving with 15 psi pressure at 123 °C for 30, 60 or 90 minutes). Since the results of these researchers differed, in this experiment we adopted treatment with CaO, diluted in water (1:10), at a dose of 60 g/kg of bran, with rest for 18 hours, which proved to be effective in detoxifying ricin.

The lambs were distributed in a randomized block design, where the block was the day of entry into the experiment, in five treatments, as follows:

- Treatment Without - Without supplementation in *creep-feeding* ;
- Treatment 0 - *Creep-feeding* supplementation, concentrate without the addition of detoxified castor bean meal (FMD);
- Treatment 33 - *Creep-feeding* supplementation, replacement of 33% soybean meal in the concentrate with FMD (based on DM);
- Treatment 67 - *Creep-feeding* supplementation, replacing 67% soybean meal in the concentrate with FMD (based on DM);
- Treatment 100 - *Creep-feeding* supplementation, with 100% soybean meal replaced by FMD in the concentrate (based on DM).

After birth, the navel was cut and disinfected with 10% iodine solution and, immediately after birth, the lambs' colostrum intake was observed.

From the tenth day of life, the animals began the period of adaptation to a solid diet, offered in *creep-feeding*. The experimental period began on the 15th day of the lambs' lives, when they were fed the diets that made up their respective treatments. The experimental period ended when the lambs reached 75 days of age. At this point, they were slaughtered to collect additional data for further analysis.

Periodic monitoring of parasitic infections was carried out by collecting feces directly from the rectal ampulla of the lambs and ewes to count the number of eggs per gram of feces (OPG), according to the methodology of Matos & Matos (1998). When the egg count per gram of feces indicated a value above 500, as referenced in the literature (Garcia, 2002), treatment with a broad-spectrum dewormer was carried out.

Concentrate (Table 1) was supplied *ad libitum* to the lambs in treatments 0, 33, 67 and 100, and was given daily. The following day, the leftovers were weighed and a new supply of concentrate was given. The amount supplied was always increased by 10% of the previous day's consumption, in order to ensure that consumption was *ad libitum*. Each of the treatments had a *creepfeeding* area, sized to allow simultaneous access for all the lambs. The management was that of a semi-confinement, where the lambs, at night, were collected, by treatment, in covered stalls, arranged on a bedding-type floor, remaining separated from their mothers until the following morning. At dawn, all the lambs and ewes were released into a single paddock. The animals were managed in two 3.5 ha paddocks on a rotational basis during the experiment, according to the availability of forage. The sheep's diet consisted of pasture with a predominance of *Brachiaria decumbens*, and they were offered standard concentrate supplementation, which consisted of cornmeal, soybean meal and mineral mix, with 14% CP.

Table 1 - Percentage levels of concentrate ingredients, based on natural matter

	Level of ingredients in the diets (%)			
Ingredients	T reatment 0	Treatment 33	TREATMENT 67	Treatment 100
Ground corn	65,84	65,84	66,03	66,03
Soybean meal	32,02	21,41	10,60	-
FMD[1]	-	10,01	20,01	30,01
Urea	-	0,55	1,09	1,64
Ammonium sulphate	-	0,06	0,12	0,18
Mineral supplement	2,14	2,14	2,14	2,14

[1] Detoxified castor bean bran;[2] Components of the mineral supplement: bicalcium phosphate; calcarlo; sodium chloride; sulphur flower; zinc sulphate; copper sulphate; cobalt sulphate; manganese sulphate; potassium iodate; sodium selenite. The mineral mixture was calculated to meet 100% of the macro and micromineral requirements.

Water was available for the animals to drink from collective troughs, which were different for the ewes and lambs. Mineral salt was provided in troughs, *ad libitum*, also separately for both the lambs and their mothers.

The lambs were weighed on the 14th day of life, before the start of the experimental period, and every two weeks until slaughter to monitor performance. For the purposes of statistical analysis of performance, only the weights at the beginning and end of the experiment were used.

2.4 - Sample processing and laboratory analysis

Samples of the leftovers, if present, were taken daily, packed in plastic bags and then frozen. Samples of the ingredients used in each concentrate preparation were taken and stored in plastic bags and frozen for later analysis. Monthly pasture samples were taken using the simulated grazing technique, in which the lambs' grazing behavior was previously analyzed and then reproduced, always by the same trained individual. The pasture samples from the simulated grazing were packed in plastic bags and frozen immediately for later analysis. The daily leftover samples were grouped proportionally, every seven days, to make up one composite sample per treatment. The samples were then partially dehydrated in a ventilated oven at 55 °C for 72 hours, ground in a knife mill with a 1 mm sieve, packed in a jar with a lid and stored for future analysis.

The chemical analyses carried out included determining the levels of dry matter, organic matter, mineral matter, ether extract (EE), total nitrogen, insoluble nitrogen in neutral detergent, neutral detergent fiber corrected for ash and protein (NDFp), lignin, calcium and phosphorus in the feed and leftovers and were carried out according to the specifications described in Detmann (2010), with crude protein (CP) being obtained by the product between the total nitrogen content and the factor 6.25. The mineral solution for determining the mineral macroelements was prepared wet (INCT-CA M-004/1 method). After the appropriate dilutions, the calcium content was determined using an atomic absorption spectrophotometer and the phosphorus content by colorimetry.

The non-protein nitrogen (NNP) of the food was determined using the trichloroacetic acid method (TCA - INCT-CA Method N-002/1).

Total carbohydrates (TC) were obtained using the equation from Sniffen et al. (1992): TC = 100 - (%PB + %EE + %Ashes).

The contents of non-fibrous carbohydrates corrected for ash and protein (NFCp) were calculated as proposed by Hall (2001), adding the correction of the NDF for ash:

CNFcp = 100 - [(%PB - %PBurea) + % urea] - %FDNcp - %EE - %Azas.

Total digestible nutrients (NDT) were calculated according to Weiss (1999), but correcting the NDF and NFC for ash and protein, using the following equation:

NDT (%) = PBD + FDNcpD + CNFcpD + 2.25EED,

where: PBD = digestible PB; FDNcpD= digestible FDNcp; CNFcpD= digestible CNFcp; and EED= digestible EE.

2.5 - Consumption and apparent digestibility of dry matter and dietary nutrients

The consumption and digestibility of the diets was assessed using the same 43 lambs as in the productive performance experiment. The area used for the animals in the digestibility trial was the same as that used in the productive performance trial.

The digestibility trial began on the 30th experimental day and lasted 13 days: seven days for adapting the animals to the chromic oxide and six more days for collecting feces, at alternate times (4 pm, 2 pm, 12 pm, 10 am, 8 am and 6 am), thus obtaining representative samples of feces from each animal. The indicator was supplied from the 1st to the 12th day of the digestibility test. The amount administered was two grams to the lambs, in a single daily dose. The chromic oxide was packaged in paper cartridges and introduced into the esophagus using an applicator.

On the seventh day of the digestibility test, pasture samples were taken from the paddocks using the simulated grazing method (Johnson, 1978). The feces was collected directly from the rectum, or immediately after the animals had defecated, without it coming into contact with the ground or any other surface other than the plastic bags used for collection. Feces samples of approximately 50 grams were collected. The pasture and feces samples were then dried in an oven with forced air ventilation at 55 °C for 72 hours. After drying, part of each sample was ground in a knife mill with a 1 mm mesh sieve, and the other part was ground in a knife mill with a 2 mm mesh sieve, and then stored as composite samples from the collection days for each animal.

Fecal dry matter excretion was estimated using the external indicator chromic oxide (Burns et al., 1994), based on the ratio between the amount of the indicator supplied and its concentration in the feces:

Fecal dry matter (g/day) = [amount of indicator supplied (g)/concentration of indicator in feces (%)] × 100

Voluntary consumption was estimated using indigestible neutral detergent fiber (iNDF) as an internal indicator, adapting the equation proposed by Detmann et al. (2001).

The total dry matter consumption of the lambs was obtained by adding up the consumption of pasture, supplement and milk produced by their respective mothers. Milk consumption was obtained five days before and five days after the start of the digestibility trial, according to the methodology used by Porto et al. (2009), adapted to the times of the experiment. The lambs were separated from their mothers for a period of twelve hours (from 6 pm to 6 am). After this separation period, the lambs were weighed and then placed in contact with the ewes to be suckled. After 30 minutes of feeding, the lambs were weighed again and milk consumption was obtained from the difference in weight between the lambs before and after feeding. To obtain the daily (24-hour) milk consumption, the difference in weight found before and after feeding was multiplied by two.

The sheep's milk was sampled and analyzed for fat, protein, lactose and total solids (Table 2). When compared, except for the higher fat content in Treatment 33, which led to an increase in total solids in this treatment, the other levels did not differ. The levels of fat, protein, lactose and total solids were similar to those found by Alves et al. (2011) for whole milk from Santa Inês sheep. They were also similar to the composition of milk from Bergamàcia sheep (Sà et al., 2005).

Table 2 - Composition of milk ingested by lambs supplemented or not with concentrate containing detoxified castor bean meal during the suckling phase by treatment

T reatment[1]	Without	0	33	67	100
Fat	5,12	5,31	7,17	5,61	5,15
Protein	5,51	5,91	5,85	5,61	5,81
Lactose	4,98	4,55	4,81	4,56	4,69
Total Solids	16,12	16,45	18,42	16,10	16,34

[1]Treatment no = no *creep-feeding* supplementation; Treatment 0: *creep-feeding* supplementation, concentrate without FMD; Treatment 33: *creep-feeding* supplementation, replacement of 33% soybean meal with detoxified castor bean meal; Treatment 67: *Creep-feeding* supplementation, substitution of 67 % soybean meal with detoxified castor bean meal; Treatment 100: *Creep-feeding* supplementation, substitution of 100 % soybean meal with detoxified castor bean meal.

Two days after estimating the lambs' milk consumption, i.e. three days before and three days after the digestibility test, milk samples were taken from all the ewes to analyze the total dry matter, protein, fat and lactose contents. The samples were analyzed in the Animal Nutrition Laboratory of the Department of Animal Science at UFV. The average levels obtained were used to estimate nutrient intake by the lambs.

2.6 - Animal slaughter and carcass characteristics

The lambs were slaughtered when they reached the age of 75 days, which in this study was taken as the age at weaning. In the 16 hours prior to slaughter, the lambs were fasted. The water diet began 16 hours before the lambs were slaughtered. Immediately before slaughter, the animals were weighed to determine their live weight at slaughter (LWW). The information related to slaughter and deemed relevant was recorded on individual cards. The animals were stunned and then slaughtered by sectioning the jugular veins and carotid arteries for total blood collection, after which they were skinned and decapitated.

The pre-stomach, abomasum and intestines were compartmentalized using double ligatures placed proximally and distally to them. Taking the double ligatures into account, the digestive tract compartments were isolated by cross-sectioning their walls and weighed individually. They were then emptied and weighed again.

The weights of blood, head, skin, feet, tail, empty viscera (rumen- reticulum, omasum, abomasum, small intestine and large intestine), mesentery, internal fat, organs (liver, heart, kidneys, spleen, lungs), and others (tongue, esophagus, trachea and reproductive system) were recorded and added to the hot carcass weight (HCW) to obtain the empty body weight (UWW).

In this study, the carcass was considered to be the final product of slaughtering the lambs, after bleeding, skinning, evisceration and amputation (head, feet, tail and testicles). After slaughter, the carcass was weighed to obtain the hot carcass weight (HCW). The carcass was then placed in a cold room at 4 °C for 24 hours. After this period, the carcass was weighed again to obtain the cold carcass weight (CWW).

Once the above-mentioned weights had been obtained, the true carcass yield (RV), commercial carcass yield (RC), biological carcass yield (RB) and weight loss due to cooling (PPR) were calculated. The following calculations were carried out (Cézar & Souza, 2007):

RV = (PCQ/PVA) x 100;

RC = (PCF/PVA) x 100;

RB = (PCQ/PCVZ) x 100;

PPR = (PCQ - PCF)/PCQ x 100.

2.7 - Isolation of the digestive tract compartments

In order to verify the proportions of the stomach and intestine compartments during the animals' development, the following ratios were used:

Digesta = weight of full viscera - weight of empty viscera;

Viscera = Rúmen-Reticulo + Omaso + Abomaso + Intestino delgado + Intestino grosso;

Stomach = Rùmen-Reticulo + Omaso + Abomaso.

Finally, measurements were taken of each compartment of the stomach, related to the PCVZ, the weight of the stomach and the weight of the viscera.

2.8 - Histological study

The ventral sac of the rumen and the distal portion of the small intestine (ileum) of the lambs were analyzed, and fragments of approximately one cm^2 were taken from these regions. The fragments were taken and fixed in 4% paraformaldehyde in 0.1 M sodium phosphate buffer, pH 7.4, for 24 hours. The fixing solution was then replaced with 0.1 M sodium phosphate buffer, pH 7.4, and the vials were stored at 4 °C (Silva, 2004).

The fragments were dehydrated in an increasing ethanol series (70% to absolute) and embedded in paraplast-type histological paraffin. Cross-sections, 5 µm thick, were obtained on a manual rotary microtome (CUT 4055, Olympus) using high-profile steel knives.

The following analyses were carried out: mast cell and intraepithelial globular lymphocyte count (modified from Bancroft & Stevens, 1996); eosinophil count (Cardia, 2009); goblet cells, acid and neutral mucins (modified from McManus & Mowry, 1960).

Images of the histological sections were captured directly from the trinocular light microscope (Olympus BX 50) by a 1.3MP CMOS camera (BioCAM) using TSViewer software.

The height of the epithelium was measured in four regions, in eight different sections, with a minimum distance of 100 µm between them, for a total of 32 papillae per animal. The images used were obtained with a 20X objective.

The above cell types were counted manually in the Image J program, using the Grid and Cell Counter tools, using a grid of 106.67x85.33 µm points, chosen at random, with a minimum distance of 100 µm between them, by microphotography.

The counts for each histological slide were grouped by animal and their averages were converted from cells/pixel to cells/mm^2 , using a scale (5.56 pixel/mm) set at 20X magnification.

The presence of *Eimeria* was also assessed using a system of crosses, which were then converted into quantifiable numbers in order to carry out statistical analysis of the data: + = 1, ++ = 2 and +++ = 3.

2.9 - Synthesis of rumen microbial nitrogen compounds through urinary excretion of purine derivatives

Microbial protein production was estimated using purine bases as a microbial indicator, which were quantified according to the technique of Ushida et al. (1985). *Spot* urine was collected from the 43 lambs used in the digestibility trial during spontaneous urination at 3pm on the 42nd day of the experimental period.

After homogenization and filtering, 10 mL aliquots were collected from the urine and diluted in 40 mL of 0.036 N sulphuric acid, as described by Valadares et al. (1999).

The pH of these samples was adjusted to below 3 to avoid bacterial destruction of the purine derivatives and the samples were then packed in plastic containers, duly identified, and frozen for later analysis of urea, total nitrogen, creatinine, xanthine, hypoxanthine, uric acid and allantoin.

The total daily urinary volume was estimated by dividing the daily urinary excretion of creatinine by the observed values of creatinine concentration in the urine, according to Valadares Filho & Valadares (2001).

Creatinine was estimated using the modified diacetyl method (commercial *kit*) and used to calculate the estimated urinary volume, based on the value of daily creatinine excretion (Y, mg/kg BW) estimated by the average daily creatinine excretion (14.25 mg/kg BW) for sheep (Pereira, 2012). The average body weight used in the calculation was obtained from five weighings, with an interval of 15 days between them. Determinations of uric acid and urea were carried out using commercial *kits* (Labtest).

Analyses of purine derivatives (xanthine, hypoxanthine, allantoin and uric acid) were carried out using the colorimetric method, according to the technique of Fujihara et al. (1987), modified by Chen & Gomes (1992). The total excretion of purine derivatives was calculated by adding up the amounts of xanthine, hypoxanthine, allantoin and uric acid excreted in the urine, expressed in mmol/day.

The microbial purines absorbed (X, mmol/day) were calculated from the excretion of purine derivatives (Y, mmol/day), using the equation $Y = 0.85X + 0.385\ PC^{0,75}$, where 0.85 represents the recovery of purines absorbed as urinary purine derivatives and $0.385\ PC^{0,75}$ the endogenous contribution to purine excretion (Verbic et al., 1990). The intestinal flow of microbial nitrogen compounds (N) (Y, g N/day) was calculated as a function of the microbial purines absorbed (X, mmol/day), using the equation $Y = (70X)/(0.83 \times 0.116 \times 1000)$, where 70 represents the N in the purines (mg N/mmol); 0.83, the digestibility of microbial purines and 0.116, the N-RNA ratio:NTotal in bacteria (Chen & Gomes, 1992).

2.10 - Efficiency of utilization of nitrogen compounds: urea nitrogen in the blood and fecal and urinary nitrogen excretion

Blood samples were taken from the 43 lambs used in the digestibility test on the 42nd day of the experiment by puncturing the jugular vein using vacuum test tubes with gel separators. Immediately, they were centrifuged at 5,000 rpm for 15 minutes, and serum samples were then taken, stored in glass containers and frozen at -15 °C for subsequent determination of the concentration of urea nitrogen and the liver enzymes aspartate aminotransferase, alanine aminotransferase and gamma glutamyltransferase.

2.11 - Nitrogen balance

The balance of nitrogen compounds was obtained from the difference between the total nitrogen ingested (N-total) and the total nitrogen excreted in feces (N-feces) and urine (N-urine). Total nitrogen in feces and urine was determined according to the methodology described in Silva & Queiroz (2002).

2.12 - Statistical analysis

The statistical model used was: $Yijk = \mu + Ti + Bj + eijk$, where:

Yij = experimental response measured under treatment i, in repetition k of block j, where K >1;

μ = overall average;

Ti = effect of treatments (described above);

Bj = block effect (age at entry into the experiment);

Eij = random error.

The sum of squares of the treatments was broken down using orthogonal contrasts, according to the scheme described in Table 3.

Table 3 - Distribution of the coefficients used in the contrasts

Contrast[1]	T reatment				
	Without	0	33	67	100
CO	4	-1	-1	-1	-1
Linear	0	-3	-1	1	3
Quadratic	0	-1	1	1	-1

Cube	0	-1	3	-3	1

[1] CO = Control *versus* supplements; Linear, Quadratic and Cube effects refer to the replacement of soybean meal with castor bean meal.[2] Treatment no = no *creep-feeding* supplementation; Treatment 0: *creep-feeding* supplementation, concentrate without castor bean meal; Treatment 33: *creep-feeding* supplementation, replacement of 33% soybean meal with detoxified castor bean meal; Treatment 67: *Creep-feeding* supplementation, 67% soybean meal replaced by detoxified castor bean meal; Treatment 100: *Creep-feeding* supplementation, 100% soybean meal replaced by detoxified castor bean meal.

The statistical procedures, contrast, mean and standard error of the mean, were carried out using the Mixed Procedure program (PROC MIXED), using the statistical package SAS (*Statistical Analysis System 9.3*), adopting 0.10 as the critical probability level for type I error. The Kruskal-Waliis H-test was used for the non-parametric analysis of the *Eimeria* spp. count variable in the dorsal sac of the rumen and ileum, converting the elements of the cross system into quantitative numbers, using the online statistical software VassarStats.

RESULTS AND DISCUSSION

The chemical composition of the ingredients in the experimental diets is shown in Table 4. The first pasture sample differed from the second, being less fibrous, NDF and NDFi. It is suggested that this was due to the drought between the two sampling dates. Soybean meal was higher in protein than castor bean meal. The ingredients also differed in terms of MM and NDF. The higher MM in FMD is possibly due to the alkaline treatment used for detoxification. As for NDF, it is believed that the high shell content of castor bean bran increased the insoluble fraction of the diet. Due to the nature of the bran, an agro-industrial co-product, its composition is heterogeneous, but it tends to have high levels of insoluble compounds, which can be seen from the levels of NDF.

In Table 5, replacing soybean meal with FMD led to a gradual increase in MM, NDF and Ca, due to the high amount of hulls in the meal and the detoxification method used, as mentioned above. It should be noted, however, that NDF was higher in Treatment 0, gradually decreasing. Its interpretation becomes clearer when expressed as a ratio of NDFi/NDF, which increases substantially from Treatment 0 to 100. The proportion of ingredients in the diets, on average, did not lead to major differences in the chemical composition of the feed.

Table 4 - Chemical composition of the ingredients used in the diet

Item (%)	Pasture 1[1]	Pasture 22	Corn	F. Soy	FMD	Urea/AS	Mineral supplement
MS	20,60	30,06	88,39	89,60	89,77	97,6	97,18
MM	7,29	7,78	1,26	5,40	9,42	-	-
PB	18,53	11,09	9,40	51,00	35,78	265,78	-
EE	2,36	1,45	0,53	2,22	0,38	-	-
NDF	63,07	69,67	53,12	67,48	63,51		
CpNDF	41,56	52,87	35,63	28,60	35,61	-	-
FDNi	12,62	21,18	2,32	1,73	42,61	-	-
Lignin	-	8,05	-	-		-	-
CNF	30,26	26,81	53,18	12,79	18,81	-	-
PIDN	18,69	12,88	10,71	34,14	22,36	-	-
CIDN	2,81	3,92	6,78	4,73	5,54	-	-

NNP	-	6,32	6,25	34,56	24,23	-	-
Ca	0,33	0,49	0,03	0,39	2,03	-	-

[1] = First harvest of *Brachiaria decumbens* pasture;[2] = First harvest of *Brachiaria decumbens* pasture.

Table 5 - Proportion of ingredients and chemical composition of the diets obtained

Ingredients	0	33	67	100
		Proportion (%)		
Corn	65,84	65,84	66,03	66,03
FS	32,02	21,41	10,60	-
FMD	-	10,01	20,01	30,01
Urea	-	0,55	1,09	1,64
S.A.	-	0,06	0,12	0,18
MM	2,14	2,14	2,14	2,14
	Nutritional composition (%)			
MS	88,97	89,03	89,09	89,16
MM	2,56	2,93	3,29	3,66
PB	22,52	22,30	22,00	21,79
PIDN/PB	79,86	74,43	68,93	63,25
EE	1,06	0,86	0,66	0,47
CT	73,86	73,91	74,05	74,09
NDF	56,58	55,77	54,94	54,13
CpNDF	32,62	33,15	33,69	34,21
FDNi	2,08	6,16	10,24	14,32
FDNi/FDN	3,68	11,05	18,64	26,45
CNFcp	41,25	41,17	41,17	41,10
PIDN	17,98	16,60	15,17	13,78
CIDN	5,98	6,03	6,08	6,14
NNP	15,18	13,94	12,64	11,40
Ca	0,14	0,31	0,47	0,63
NDT	91,76	91,55	91,35	91,14

The intakes of DM, CP, EE and NDF are shown in Table 6. CMS and CPB were higher in the treatments that received concentrate, and the contrast for these characteristics was significant. Milk consumption (SMMilk) did not differ between treatments. This implies that the differences in performance were due to the consumption of the other feedstuffs offered in the experiment. There was a linear and quadratic effect on concentrate consumption (CMSconc), so that the inclusion of castor bean meal in the concentrate, as a substitute for soybean meal, led to a reduction in CMSconc. However, this implied an increase in pasture intake (CMSpasture), which is evidenced by the CMS, where only the consumption of Treatment Without was lower than the others. CMSpasto showed a linear effect, but inversely to CMSconc, the inclusion of castor bean meal led to an increase in CMSpasto. EEC and CFDN did not differ statistically between the different treatments. The consumption of the various nutritional entities was lower than that reported by various authors (Santos, 2011; Manera et al., 2014; Morgado et al., 2014; Santos et al., 2014). It is believed that this difference was due to the category worked on. Both are presented as the calf category, but in the experiments mentioned above, the animals came in heavier and older, naturally showing higher consumption. Silva (2009), when working with confined lambs from the 14th day onwards, receiving bran or pelleted feed in *creep-feeding*, found an average daily feed consumption of 210 g for males. Mendes et. al. (2003), evaluating the performance of suckling Santa Inês lambs supplemented *with creep-feeding,* found an average feed consumption of 127 g.

The digestibility of the nutritional fractions is shown in Table 7. The averages did not differ between treatments, except for DMF, which showed a linear effect (P=0.045), and DFDN, whose significant effect was quadratic (P=0.062). The DMS and DFDN were higher than those reported in the literature (Rodrigues et al., 2008; Santos, 2011; Morgado et al., 2014). The high DMF is possibly due to the lambs being suckled, since milk has excellent digestibility. The low DFDN can be explained by rumen development, which is still incipient at this stage of growth, when fiber digestibility is compromised by the low fermentative capacity of the rumen (Mirza & Provenza, 1990; Thorhallsdottir et al., 1990). The DPB and DEE were similar to those found by Santos et al. (2014), but higher than those reported by Rodrigues et al. (2008) and Santos (2011), in the latter case only for DPB, since DEE was not reported in the work.

PI (Table 8) did not differ between treatments, which indicates good uniformity of the animals at the start of the experiment. The GP was significant for the control and quadratic and cubic effects. The quadratic effect can be understood as the concentration of the highest intakes in the medium treatments. As for the cubic effect, it is believed that this occurred because Treatment 33 was more affected than the others by *Eimeria*, which led to lower weight gain (Berriatua et al., 1994). The compromised gain of Treatment 33 generated flexibility in the described parabola, whose mathematical function, in this case, resembles a cubic curve. GMD was significant for the control and showed a quadratic and cubic effect. The results were within the range suggested by the NRC (1985) as normal, from 50 to 100 g/day to 350 to 400 g/day. GMD was similar to that found by Poli et al. (2008) for weaned lambs, but was, on average, lower than those reported in the literature (Santos et al., 2014; Zundt et al., 2014; Rodrigues et al., 2008; Carvalho et al., 2007), which ranged from 112 to 277 g/day. However, this small difference between the values is possibly associated with the different experimental diets and ages of the animals in the study. The AE was significant for the cube effect. The biological understanding of this contrast is the same as for the GP and GMD variables, since EE is a function of gain through consumption.

Carcass yields (Table 9) did not differ between treatments, except for a cyclical effect on commercial yield. Carcass yields were similar to those reported by França (2011), Santos (2011) and Ribeiro et al. (2009), but higher than those suggested by Oliveira et al. (2014). When compared with other literature, the carcass yields were similar to those observed by Cândido et al. (2008), working with the substitution of soybean meal by detoxified castor bean meal at four levels (0.0, 33.0, 67.0 and 100.0%), which were true carcass yield (43.31%, 43.03%, 42.16% and 45.10%), commercial (42.96%, 42.57%, 41.75% and 44.08%) and biological (56.10%, 55.05%, 53.63% and 57.56%), respectively. The results are also similar to those observed by Gonzaga Neto et al. (2005), where the results for true, commercial and biological carcass yields in Santa Inês x Dorper sheep were 51.4%, 50.5% and 62.2% respectively. Cooling loss showed a quadratic effect, being higher in the treatments without castor beans. This is possibly associated with the deposition profile in the carcass. Animals with higher fat cover have lower chilling loss (Rodrigues et al., 2008). The PPR averages were similar to those reported by Oliveira et al. (2014), Santos (2011) and Gomes et al. (2010).

The urinary volume (Table 10) and the levels of excretion of xanthine-hypoxanthine, uric acid and purine derivatives, as well as the levels of purines absorbed, showed no significant effect for any of the contrasts. AU showed the lowest average in Treatment 33. This is possibly associated with protein deposition, reflected in the lower weight gain by the animals in this treatment. As the CPB in this treatment was similar to the others, it is believed that the protein was mobilized for other purposes, as a response to the eimeriosis. The demand for amino acids increases when animals are subclinically infected, due to the need to repair damaged tissues and the rapid immune response. The loss in the supply of amino acids to other tissues significantly reduces anabolism, compromising production indices (Cardia et. al., 2009). Urinary allantoin and urea excretion had a significant quadratic and linear effect, respectively. Urinary volume was similar to that found by Pereira (2012). The levels of UA, XH and DP were different from those reported by Pereira (2012) and Oliveira (2010), which are different from each other. The levels of A and PUR were similar to those suggested by the two authors. The proportion of purines (A, AU and XH) in relation to the total purines excreted (DP) was 49.1:1.2:49.1, respectively. This ratio was different to that reported by Pereira (2012), of 14.3:7.1:78.6, Oliveira (2010), of 85.2:10.4:4.40, and Chen & Gomes (1992). Urinary urea levels were much higher than those recommended by Kaneko et al. (2008), from 0 to 1.9 mg/dL for this category. However, urinary urea excretion was lower than that reported by Pereira (2012) and Santos (2011) and similar to that suggested by Oliveira (2010).

The intestinal flow of microbial nitrogen compounds (Table 11) and the total N retained were not significant for any of the contrasts. However, the total N ingested and absorbed, as well as the efficiency of N ingested and absorbed were, respectively, significant for control (P=0.020 and P=0.045) and linear effect (P=0.008 and P=0.009). The flow of microbial N was lower than that found by Pereira (2012). This is possibly due to the low rumen fermentation capacity of the lambs used in this study. The levels of N ingested, absorbed and retained were lower than those reported in the literature (Furtado et al., 2014; Morgado el al., 2014, Pereira, 2012; Santos, 2011; Oliveira, 2010). It is believed that this is associated with the age of the lambs, which were slaughtered in the transition between the breeding and rearing phases. The efficiency of N ingested and absorbed, as a function of N retained, was, respectively, higher and lower than that found by Morgado et al. (2014).

Liver condition was analyzed based on serum levels of aspartate aminotransferase (AST), alanine aminotransferase (ALT) and gamma gultamyltransferase (GGT) (Table 12), since these are indicators of liver function, which are elevated in the event of liver damage and are associated with the occurrence of ricin poisoning in rats and sheep (Kumar et al., 2003; Aslani et al., 2007). Serum urea, AST and ALT levels were significant for quadratic (P=0.039) and cubic (P=0.060) effects for the former, linear (P=0.070) for the latter and quadratic (P=0.096) for the latter. GGT levels did not differ for any of the contrasts. Serum urea levels are above the levels found by Kaneko et al. (2008), from 8 to 20 mg/dL. Serum AST levels were within the range found by Radostists et al. (2007), from 60 to 280 IU/L, and by Kaneko et al. (2008), from 0 to 90 IU/L. AST levels were similar to those suggested by Sousa (2014) and Santos (2011), but lower than those reported by Oliveira (2012). Serum ALT levels were low when compared to those found by Radostists et al. (2007), from 22 to 28 IU/L, but normal when compared to the levels found by Kaneko et al. (2008), from 0 to 30 IU/L. ALT levels were lower than those found by Oliveira (2012) and Santos (2011). GGT levels were similar to those described by Sousa (2014).

The length of the small intestine (Table 13) showed no significant effect for any contrast. Analyzing each component of the small intestine separately, the lengths of the duodenum and ileum showed a quadratic effect (respectively, P=0.069 and P=0.036), while that of the jejunum was only significant for a cubic effect (P=0.020). Values for the length of the small intestine and its parts have not been reported for such young lambs.

The number of eosinophils (Table 14) showed a linear effect, indicating a tendency for the body's defense response to increase as the proportion of castor bean bran in the diet increased. Treatments 0 and 66 had the lowest averages, possibly due to higher dry matter consumption. Animals whose demands are properly met tend to show a better response to allergens and/or infectious agents. The number of eosinophils in the ileum of the lambs, per mm^2 , was significantly lower than that described by Cardia et al. (2009) for the duodenum and jejunum of lambs fed *ad libitum*, in the Control group. The eosinophil undergoes complex modifications so that it can migrate from the circulation to the tissues, thus becoming a complex multifunctional cell, acting both in inflammation with cytotoxic functions and in the regulatory action of the tissue inflammatory response, especially immunity (Couissinier-Paris, 2006).

The number of mast cells-lymphocytes (Table 14) did not differ statistically between treatments. The averages were much lower than those reported by Cardia et al. (2009). The type of cell infiltrating the lamina propria defines the type of inflammatory reaction, which can be eosinophilic or lymphoplasmocytic, depending on the region of the GIT (I lall, 1999). In this study, the number of eosinophils was substantially higher than that of mast cells, which would characterize an eosinophilic chronic inflammatory bowel disease (IBD).

Mucins are present on the cell surface of different tissues, performing, among other things, a protective function. They are also involved in biological processes such as renewal, differentiation, signaling, infection and inflammation (Winterfeld et al., 2003; Andrianifahanana et al., 2006; Seeberger & Wernz, 2007). Mucins are easily stained by PAS and alcian blue. PAS stains neutral mucins, due to the richness of polysaccharides inside these cells. AB staining stains acid mucins, staining the mucins blue and the nuclei red. PAS-positive, AB-positive and PAS/AB-positive cells were not significant for any of the contrasts. Comparative data was not found, as histological analyses of neutral and acidic mucins are not common in sheep experiments.

The smallest and largest epithelium thicknesses (Table 15) of the papillae of the ventral rumen sac were significant for the control (respectively, P=0.002 and P=0.084) and cube effect (respectively, P=0.018 and P=0.010). The No Treatment had higher averages for the smallest and largest thicknesses than the others. This is possibly due to a higher fibre intake, since the lambs in this treatment met all their demand from milk and pasture intake, thus favouring greater keratinization of the rumen, which in this case is expressed by the thickness of the epithelium.

The presence of *Eimeria* spp. (Table 16) indicates the degree of parasitic infestation of the lambs. The presence of *Eimeria* was significant in both the kidneys (H=5.10) and the ileum (H=7.03). Coccidiosis, a disease caused by protozoa of the genus Eimeria, becomes more important when farming is intensified, where the consumption of concentrate is an important factor, aggravating the infection (Catchpole et al., 1993; Urquhart et al., 1998). During the study, calves and dams from the different treatments grazed at the same time in the same paddock. Gastrointestinal parasites mainly affect lambs and, even when subclinical, can cause a reduction in weight gain efficiency, as well as increasing the animals' susceptibility to other diseases (Foreyt, 1990; Daugschies et al., 1997; Faizal et al., 1999; Svensson, 2000; Daugschies & Najdrowski, 2005; Joley & Bardsley, 2006; Sotomaior et al. 2009). Contamination occurs through the ingestion of sporulated oocysts present in the environment, often in animals under one year of age, whose immunity is still developing (Lima, 2004; Mitchell et al., 2012; Andrews, 2013). The Kruskal-Wallis H-test indicates statistical significance, but as it is a non-parametric test, it does not indicate where the difference lies. It is believed that Treatment 33, as it showed worse results in terms of animal performance, was more affected by the *Eimeria* spp. infection.

Table 6 - Consumption of dry matter (DM), crude protein (CPB), ethereal extract (EEC) and neutral detergent fiber (NDF) of lambs supplemented or not with concentrate containing detoxified castor bean bran during the suckling phase

Immediately[1]						Contrast (P-value)[2]			
Variable (g/day)	Without	0	33	67	100	CO	L	Q	C
	276	+321	330	+326	+352	+			
CMS	22,7	+22,7	24,0	22,7	24,0	0,033	0,403	0,706	0,684
	24,4	+33,8	+30,5	+35,8	+				
CMSleite	9,47	9,36	9,77	9,31	39,7+1,01	0,214	0,509	0,643	0,777
CMSconc	0	95,9	58,9	34,1	28,6	0,0001	0,0001	0,0001	0,2955
		230+	238+	256+					
CMSpasto	249+ 19,5	13,0	12,2	11,5	280+ 12,2	0,933	0,069	0,706	0,977
	38,4	+51,9	+50,3	+47,3	+50,3	+			
CPB	3,46	3,46	3,67	3,46	3,67	0,005	0,620	0,526	0,656
	6,25	+6,55	+7,49	+7,03	+7,73	+			
CEE	0,59	0,59	0,62	0,59	0,62	0,157	0,260	0,853	0,352
	132	+121	132	+132	+141	+			
CFDN	10,09	+9,51	10,09	9,51	10,09	0,971	0,165	0,955	0,651

[1]Treatment no = no *creep-feeding* supplementation', Treatment 0: *creep-feeding* supplementation, concentrate without castor bean meal; Treatment 33: *creep-feeding* supplementation, substitution of 33 % soybean meal with detoxified castor bean meal; Treatment 67: *Creep-feeding* supplementation, substitution of 67 % soybean meal with detoxified castor bean meal; Treatment 100: *Creep-feeding* supplementation, substitution of 100 % soybean meal with detoxified castor bean meal. [2]CO = control vs. supplements; Linear (L), Quadratic (Q) and Cube (C) effects refer to the replacement of soybean meal with detoxified castor bean meal.

Table 7 - Digestibility of dry matter (DMS)$_1$ crude protein (DPB)$_1$ ether extract (DEE) and neutral detergent fiber (DFDN) of lambs supplemented or not with concentrate containing detoxified castor bean bran during the suckling phase

Immediately[1]						Contrast (P-value)[2]			
Variable (%)	Without	0	33	67	100	CO	L	Q	C
DMS	90,4+1,66	85,9+1,66	88,5+1,77	90,8+1,66	90,4+1,77	0,441	0,045	0,397	0,754
		86,0	+89,8	+88,0	+89,4	+			
DPB	84,4 + 3,41	3,40	3,49	3,40	3,49	0,126	0,407	0,598	0,391
		80,5	+82,7	+81,5	+84,0	+			
DEE	80,4 + 3,09	3,40	3,18	3,09	3,18	0,500	0,394	0,953	0,517
		12,6	+12,0	+11,5	+12,7	+			
DFDN	11,8 + 0,61	0,61	0,55	0,52	0,58	0,402	0,894	0,062	0,426

[1]Treatment no = no *creep-feeding* supplementation', Treatment 0: *creep-feeding* supplementation, concentrate without castor bean meal; Treatment 33: *creep-feeding* supplementation, replacement of 33 % of soybean meal with detoxified castor bean meal; Treatment 67: *Creep-feeding* supplementation, substitution of 67 % soybean meal with detoxified castor bean meal; Treatment 100: *Creep-feeding* supplementation, substitution of 100 % soybean meal with detoxified castor bean meal. [2]CO = control vs. supplements; Linear (L), Quadratic (Q) and Cube (C) effects refer to the replacement of soybean meal with detoxified castor bean meal.

Table 8- Initial weight (IW), weight gain (WG), average daily gain (ADG) and feed efficiency (FE) of lambs supplemented or not with concentrate containing detoxified castor bean meal during the suckling phase

Immediately[1]						Contrast (P-value)[2]			
Variable	Without	0	33	67	100	CO	L	Q	C
PI (kg)	7,85+1,60	7,95+1,59	7,99+1,61	7,92+1,58	8,22+1,64	0,817	0,811	0,850	0,878
GP (kg)	4,85+1,90	7,56+1,89	3,83+1,91	7,67+1,89	6,36+1,93	0,053	0,941	0,087	0,0002
GMD (g/day)	80,8 + 31,6	126,1+31,5	63,8 31,8	+127,8 31,5	+106,0 32,2	+ 0,053	0,941	0,087	0,0002
AS (%)[3]	40,5+15,0	52,6+14,9	20,9+15,2	43,8+14,9	32,1 15,4	+ 0,714	0,280	0,213	0,017

[1]Treatment no = no *creep-feeding* supplementation', Treatment 0: *creep-feeding* supplementation, concentrate without castor bean meal; Treatment 33: *creep-feeding* supplementation, substitution of 33 % soybean meal with detoxified castor bean meal; Treatment 67: *Creep-feeding* supplementation, substitution of 67 % soybean meal with detoxified castor bean meal; Treatment 100: *Creep-feeding* supplementation, substitution of 100 % soybean meal with detoxified castor bean meal. [2]CO = control vs. supplements; Linear (L), Quadratic (Q) and Cube (C) effects refer to the replacement of soybean meal with detoxified castor bean meal.[3] EA = (GMD/CMS)*100.

1Table 9 - True carcass yield (RV), commercial carcass yield (RC) and biological carcass yield (RB) and chilling loss (PPR) of lambs supplemented or not with concentrate containing detoxified castor bean meal during the suckling phase

Immediately[1]						Contrast (P-value)[2]			
Variable (%)	Without	0	33	67	100	CO	L	Q	C
RV	43,6 + 2,83	44,2 2,81	+41,3 2,85	+45,8 2,80	+44,7 2,91	+ 0,771	0,333	0,497	0,034
RC	42,1+3,25	42,2 3,23	+40,2 3,27	+44,8 3,23	+ 43,1+3,33	0,777	0,239	0,925	0,042
RB	65,1 + 1,93	62,5+1,91	61,9+1,96	65,3+1,90	64,5 2,02	+ 0,229	0,077	0,962	0,121
PPR	3,51 + 1,21	5,11 1,19	+ 2,49+1,23	2,39+1,19	3,27+1,27	0,822	0,115	0,031	0,664

[1]Treatment no = no *creep-feeding* supplementation', Treatment 0: *creep-feeding* supplementation, concentrate without castor bean meal; Treatment 33: *creep-feeding* supplementation, substitution of 33 % soybean meal with detoxified castor bean meal; Treatment 67: *Creep-feeding* supplementation, substitution of 67 % soybean meal with detoxified castor bean meal; Treatment 100: *Creep-feeding* supplementation, substitution of 100 % soybean meal with detoxified castor bean meal. [2]CO = control vs. supplements; Linear (L), Quadratic (Q) and Cube (C) effects refer to the replacement of soybean meal with detoxified castor bean meal.

Table 10- Urinary volume (VU), xanthine-hypoxanthine (XH)1 allantoin (A), uric acid (AU), purine derivatives (DP), absorbed purines (PUR) and urinary urea (U) of lambs supplemented or not with concentrate containing detoxified castor bean bran during the suckling phase

Immediately[1]						Contrast (P-value)[2]			
Variable	Without	0	33	67	100	CO	L	Q	C
VU (L/day)	1,48 + 0,31	1,38 + 0,31	1,37 + 0,32	1,53 + 0,31	1,51+0,32	0,921	0,615	0,962	0,772
XH (mmol/day)	9,05+1,48	9,72 + 1,48	9,03+1,55	8,90+1,48	9,14+1,55	0,924	0,759	0,735	0,974
A (mmol/day)	8,40 + 2,56	12,92 + 2,52	7,76 + 2,65	6,40 + 2,51	9,43 + 2,76	0,755	0,226	0,064	0,952
AU (mmol/day)	0,21 + 0,02	0,22 + 0,02	0,17 + 0,02	0,20 + 0,02	0,21+0,02	0,614	0,755	0,161	0,235
SD (mmol/day)	18,0 + 3,35	22,9 + 3,33	17,2 + 3,53	15,4 + 3,32	19,4 + 3,60	0,846	0,407	0,150	0,893
PUR (mmol/day)	18,3 + 3,73	23,7 + 3,73	17,5 + 3,96	14,8 + 3,73	19,9 + 3,96	0,882	0,419	0,153	0,805
U (g/day)	10,0 + 2,60	11,4 + 2,60	10,7 + 2,67	10,2 + 2,60	6,85 + 2,67	0,908	0,102	0,484	0,719

[1]Treatment no = no *creep-feeding* supplementation', Treatment 0: *creep-feeding* supplementation, concentrate without castor bean meal; Treatment 33: *creep-feeding* supplementation, substitution of 33 % soybean meal with detoxified castor bean meal; Treatment 67: *Creep-feeding* supplementation, substitution of 67 % soybean meal with detoxified castor bean meal; Treatment 100: *Creep-feeding* supplementation, substitution of 100 % soybean meal with detoxified castor bean meal. [2]CO = control vs. supplements; Linear (L), Quadratic (Q) and Cube (C) effects refer to the replacement of soybean meal with detoxified castor bean meal.

Table 11 - Intestinal flow of microbial nitrogen compounds (Nmic)$_1$ total N ingested (Ning), absorbed (Nabs) and retained (Nret) and efficiency of N ingested (Efil) and absorbed (Efi2) of lambs supplemented or not with concentrate containing detoxified castor bean meal during the suckling phase

Immediately[1]						Contrast (P-value)[2]			
Variable	Without	0	33	67	100	CO	L	Q	C
Nmic (g of		17,2	+12,7	+10,8	+14,5	+			
N/day)	13,3 + 2,71	2,71	2,88	2,71	2,88	0,882	0,419	0,153	0,805
			8,04	+7,57	+8,04	+			
Ning (g/day)	6,70 + 0,47	8,31+0,47	0,50	0,47	3,33	0,020	0,560	0,456	0,601
			7,20	+6,73	+7,15	+			
Nabs (g/day)	6,03 + 0,45	7,21+0,45	0,47	0,45	0,47	0,045	0,759	0,639	0,513
		3,83	+3,27	+3,90	+4,78	+			
Nret (g/day)	3,12 + 0,53	0,53	0,56	0,53	0,56	0,173	0,157	0,193	0,693
		40,4	+40,5	+50,3	+59,8	+			
Efil (%Nret)	46,0 + 5,61	5,61	5,91	5,61	5,91	0,750	0,008	0,385	0,606
		45,5	+44,5	+56,8	+66,8	+			
Efi2 (%Nret)	51,4 + 5,82	5,82	6,18	5,82	6,18	0,774	0,009	0,402	0,686

[1]Treatment no = no *creep-feeding* supplementation', Treatment 0: *creep-feeding* supplementation, concentrate without castor bean meal; Treatment 33: *creep-feeding* supplementation, substitution of 33 % soybean meal with detoxified castor bean meal; Treatment 67: *Creep-feeding* supplementation, substitution of 67 % soybean meal with detoxified castor bean meal; Treatment 100: *Creep-feeding* supplementation, substitution of 100 % soybean meal with detoxified castor bean meal. [2]CO = control vs. supplements; Linear (L), Quadratic (Q) and Cube (C) effects refer to the replacement of soybean meal with detoxified castor bean meal.

Table 12- Serum urea (Userica)$_1$ aspartate aminotransferase (AST), alanine aminotransferase (ALT), gamma glutamyltransferase (GGT) of lambs supplemented or not with concentrate containing detoxified castor bean meal during the suckling phase

Immediately[1]						Contrast (P-value)[2]			
Variable	Without	0	33	67	100	CO	L	Q	C
Userica		34,2	+56,4	+37,6	+33,6	+			
(mg/dL)	32,4 + 5,94	5,91	7,72	5,90	6,32	0,209	0,496	0,039	0,060
		105	+	84,0	+81,9	+			
AST (U/L)	97,0 + 9,11	9,11	92,0+12,2	9,11	9,65	0,559	0,070	0,573	0,996
		13,4	+10,2		+				
ALT (U/L)	8,56+1,56	9,56+1,56	2,10	1,56	8,13+1,66	0,329	0,315	0,096	0,328
GGT (U/L)	61,2+13,4	55,4+13,3	78,8+17,2	53,4+13,2	45,3+14,5	0,832	0,339	0,258	0,311

[1]Treatment no = no *creep-feeding* supplementation', Treatment 0: *creep-feeding* supplementation, concentrate without castor bean meal; Treatment 33: *creep-feeding* supplementation, substitution of 33 % soybean meal with detoxified castor bean meal; Treatment 67: *Creep-feeding* supplementation, substitution of 67 % soybean meal with detoxified castor bean meal; Treatment 100: *Creep-feeding* supplementation, substitution of 100 % soybean meal with detoxified castor bean meal. [2]CO = control vs. supplements; Linear (L), Quadratic (Q) and Cube (C) effects refer to the replacement of soybean meal with detoxified castor bean meal.

1Table 13- Total length of the duodenum (DUOD), jejunum (JEJ), ileum (ILEO) and small intestine (DEL) of lambs supplemented or not with concentrate containing detoxified castor bean meal during the suckling phase.

Immediately[1]						Contrast (P-value)[2]			
Variable (cm)	Without	0	33	67	100	CO	L	Q	C
	18,4	+22,5	+17,6	+20,0	+24,3	+			
DUOD	2,87	2,86	2,98	2,86	2,98	0,322	0,480	0,069	0,624
		1351	1222	+1390+					
JEJ	1351 +101	+101	106	101	1136+106	0,330	0,371	0,895	0,020
		1026+							
ILEO	886+ 140	140	885+ 143	752 + 140	940 + 164	0,935	0,330	0,036	0,552
		2396+	1990	+					
DEL	2211 +127	126	133	2139+125	2079+ 137	0,653	0,130	0,145	0,150

[1]Treatment no = no *creep-feeding* supplementation', Treatment 0: *creep-feeding* supplementation, concentrate without castor bean meal; Treatment 33: *creep-feeding* supplementation, substitution of 33 % soybean meal with detoxified castor bean meal; Treatment 67: *Creep-feeding* supplementation, substitution of 67 % soybean meal with detoxified castor bean meal; Treatment 100: *Creep-feeding* supplementation, substitution of 100 % soybean meal with detoxified castor bean meal. [2]CO = control vs. supplements; Linear (L), Quadratic (Q) and Cube (C) effects refer to the replacement of soybean meal with detoxified castor bean meal.

Table 14- Number of eosinophils (EOS), mast cells and lymphocytes (ML), PAS positive cells (PAS), alcian blue positive cells (AB) and PAS and alcian blue positive cells (PAS/AB) in the ileum of lambs supplemented or not with concentrate containing detoxified castor bean bran during the suckling phase.

	Immediately[1]					Contrast (P-value)[2]			
Variable (cel/mm)[2]	Without	0	33	67	100	CO	L	Q	C
	39,3	+28,7	+39,8	+34,5	+45,0	+			
EOS	2,91	2,91	3,09	2,91	3,09	0,488	0,002	0,914	0,022
	4,01	+2,93	+5,68	+4,91	+4,78	+			
ML	0,92	0,92	0,97	0,92	0,97	0,590	0,266	0,136	0,331
		6,18	+4,59	+4,59	+5,79	+			
PAS	5,31+4,07	4,04	4,19	4,02	4,31	0,987	0,924	0,637	0,996
	33,3	+35,7	+28,	732,1	+32,9	+			
AB	8,61	8,61	+8,83	8,61	8,83	0,880	0,845	0,514	0,624
	10,1	+12,8	+11,9	+12,8	+10,3	+			
PAS/AB	5,19	5,19	5,39	5,19	5,39	0,706	0,731	0,852	0,794

[1]Treatment no = no *creep-feeding* supplementation', Treatment 0: *creep-feeding* supplementation, concentrate without castor bean meal; Treatment 33: *creep-feeding* supplementation, replacement of 33 % of soybean meal with detoxified castor bean meal; Treatment 67: *Creep-feeding* supplementation, substitution of 67 % soybean meal with detoxified castor bean meal; Treatment 100: *Creep-feeding* supplementation, substitution of 100 % soybean meal with detoxified castor bean meal. [2]CO = control vs. supplements; Linear (L), Quadratic (Q) and Cube (C) effects refer to the replacement of soybean meal with detoxified castor bean meal.

Table 15- Lowest (MIN) and highest (MAX) epithelium thickness of the papillae of the ventral kidney sac (mm) of lambs supplemented or not with concentrate containing detoxified castor bean meal during the suckling phase

Variable (mm)	Immediately[1]					Contrast (P-value)[2]			
	Without	0	33	67	100	CO	L	Q	C
	25,06	+19,82	+14,87	+	20,93	+			
MIN	2,13	2,13	2,19	21,01+2,13	2,19	0,002	0,183	0,127	0,018
	52,90	+46,52	+35,77	+51,55	+49,90	+			
MAX	4,17	4,12	4,34	4,10	4,51	0,084	0,118	0,217	0,010

[1]Treatment no = no *creep-feeding* supplementation', Treatment 0: *creep-feeding* supplementation, concentrate without castor bean meal; Treatment 33: *creep-feeding* supplementation, substitution of 33 % soybean meal with detoxified castor bean meal; Treatment 67: *Creep-feeding* supplementation, substitution of 67 % soybean meal with detoxified castor bean meal; Treatment 100: *Creep-feeding* supplementation, substitution of 100 % soybean meal with detoxified castor bean meal. [2]CO = control vs. supplements; Linear (L), Quadratic (Q) and Cube (C) effects refer to the replacement of soybean meal with detoxified castor bean meal.

Table 16- Presence of Eimeria in the connective tissue of the ventral kidney sac (EIM1) and ileum (EIM2) of lambs supplemented or not with concentrate containing detoxified castor bean meal during the suckling phase, by the Kruskal-Wallis H test.

Variable	Immediately[1]					H	P-value
	Without	0	33	67	100		
EIMI[2]	19,1	28,2	25,8	16,8	20,3	5,10	0,2772
EIM2[2]	29,3	21,1	21	24,1	13,5	7,03	0,1343

[1]Treatment no = no *creep-feeding* supplementation', Treatment 0: *creep-feeding* supplementation, concentrate without castor bean meal; Treatment 33: *creep-feeding* supplementation, substitution of 33 % soybean meal with detoxified castor bean meal; Treatment 67: *Creep-feeding* supplementation, substitution of 67 % soybean meal with detoxified castor bean meal; Treatment 100: *Creep-feeding* supplementation, substitution of 100 % soybean meal with detoxified castor bean meal. [2]Average ranks from the Kruskal-Wallis H-test.

CONCLUSIONS

Replacing soybean meal with detoxified castor bean meal, under the conditions of the experiment, did not significantly compromise the animals' performance.

The use of *creep-feeding* positively affects performance, with the gain of supplemented animals being higher than that of control animals.

The intake of detoxified castor bean bran leads to a linear increase in the number of eosinophils.

Concentrate consumption has a negative influence on the thickness of the papillae epithelium in the ventral sac of the rumen.

BIBLIOGRAPHICAL REFERENCES

ABDALLA, A.L.; SILVA FILHO, J.C.; GODOI, A.R. et al. Utilization of by-products from the biodiesel industry in ruminant feed. **Revista Brasileira de Zootecnia**, v.37, special supplement, p.260-258, 2008.

ALVES, A.C.; ALVES, N.G.; ASCARI, I.J. et al. Characterization of colostrum, transitional milk and whole milk from Santa Inês sheep. 2011. (Paper/Congress Presentation). In: Conbravet, 38, 2011, Florianópolis. **Proceedings...** Brazilian Congress of Veterinary Medicine, 38., 2011.

ANDREWS, A.H. Some aspects of coccidiosis in sheep and goats. **Small Ruminant Research**, v.110, p.93-95, 2013.

ANDRIANIFAHANANA M, MONIAUX N, BATRA SK. 2006. Regulation of mucin expression: mechanistic aspects and implications for cancer and inflammatory diseases. **Biochimestry and Biophysic Acta**, v.1765, p.189-222.

ASLANI, M.R.; MALEKI, M.; MOHRI, M. et al. Castor bean (*Ricinus communis*) toxicosis in a sheep flock. **Toxicon**, v.49, p.400-406, 2007.

BANCROFT, J.D.; STEVENS, A.; TURNER, D.R. **Theory and practice of histological techniques**. 3.ed. New York: Churchill, Livingstone, 1990. 726p.

BARROS, G. S. de C.; SILVA, A. P.; PONCHIO, L. A. et al. Biodiesel production costs in Brazil. **Revista de Politica Agricola**, Ano XV - n° 3 -Jul./Aug./Sept. 2006.

BERRIATUA, E.; GREEN, L.E.; MORGAN, K.L. A descriptive epidemiological study of coccidiosis in early lambing housed flocks. **Veterinary Parasitology**, v.54, p.337351, 1994.

BURNS, J.C.; POND, K.R.; FISHER, D.S. Measurement of forage intake. In: FAHEY JR., G.C. (Ed.) **Forage quality, evaluation and utilization**. Madison: American Society of Agronomy, p.494-531, 1994.

CÀNDIDO, M.J.D., VIEIRA, M.M.M.; BOMFIM, M.A.D.; et al. Carcass characteristics of sheep fed diets containing four levels of castor bean meal. In: ANNUAL MEETING OF THE BRAZILIAN SOCIETY OF ZOOTECHNICS, 45, 2008, Lavras. **Proceedings...** Lavras: SBZ, 2008.

CARDIA, D.F.F. **Immune response and pathophysiology of artificial infections by**

Trichostrongylus colubriformis in Santa Inês lambs. Botucatu, 2009. 42p. Dissertation (Master's Degree) - Faculty of Veterinary Medicine and Zootechnics, São Paulo State University.

CARVALHO, S.; BROCHIER, M.A.; PIVATO, J. et al. Weight gain, carcass characteristics and non-carcass components of Texel lambs finished on different feeding systems. **Ciência Rural**, v.37, n.3, p.821-827, 2007.

CATCHPOLE, J.; NORTON, C.C.; JOYNER, L.P. The occurrence of *Eimeria weybridgensis* and other species of coccidia in lambs in England and Wales. **British Veterinary Journal**, v.131, p.392-401, 1975.

CEZAR, M.F.; SOUSA, W.H. Sheep and goat carcasses: procurement-evaluation-classification. Uberaba: **Agropecuâria Tropical,** 2007. 232p.

CHEN, X.B.; GOMES, M.J. **Estimation of microbial protein supply to sheep and cattle based on urinary excretion of purine derivatives - an overview of technical details.** International Feed Research Unit. Aberdeen: Aberdeen Rowett Research Institute, 1992. 21p. (Occasional publication).

COUISSINIER-PARIS, P. Activated eosinophils: techniques to characterize them. **Presse Medicine**, v.35, n.1-2, p.125-134, 2006.

DAUGSCHIES, A.; BÜRGER, H-J.; AKIMARU, M. Effects of experimental infections with *Eimeria bovis* on the balance of sodium, potassium and water in calves. **Parasitology International**, v 46, p.159-169, 1997.

DAUGSCHIES, A.; NADJROWSKI, M. Eimeriosis in cattle. Current understanding. **Journal of Veterinary Medicine B**, v.52, p.417-427, 2005.

DOAN, L.G. Ricin: mechanism of toxicity, clinical manifestations, and vaccine development. **Journal of Toxicology-Clinical Toxicology**, New York, v.42, n.2, p.201-208, 2005.

DETMANN, E.; PAULINO, M.F.; ZERVOUDAKIS, J.T. et al. Chromium and internal indicators in the estimation of intake of supplemented mestizo steers at pasture. **Revista Brasileira de Zootecnia**, v.30, n.5, p.1600-1609, 2001.

DETMANN, E. Fiber in the nutrition of dairy heifers In: PEREIRA, E.S. et al. Dairy **Heifers.** Fortaleza: Graphiti Gràfica e Editora Ltda, 2010. Chap.8, p. 253332.

EVANGELISTA, A.R.; ABREU, J.G.; PERON, A.J. Evaluation of the chemical composition of castor and peanut cake obtained by different oil extraction methods. In: Brazilian Castor Bean Congress, Campina Grande, 2004. **Proceedings...** Campina Grande: Brazilian Agricultural Research Corporation. Available at: www.cnpa.embrapa.br.

FAIZAL, A.C. M.; RAJAPAKSE, R.P.V.J.; JAYASINGHE, S.R. et al. Prevalence of *Eimeria* spp. and gastrointestinal nematodes versus weight gains in treated goats raised in the dry areas of Sri Lanka. **Small Ruminant Research**, v.34, p.21-25, 1999.

FERREIRA, G.B.; BERTRAÂO, N.E.N.; SEVERINO, L.S. et al. **Castor bean cultivation in the cerrado: risks and opportunities.** Embrapa Algodao: Campina Grande, 70p., 2006 (Documentos, 149).

FOREYT, W.J. Coccidiosis and cryptosporidiosis in sheep and goats. **Veterinary Clinics of North America - Food Animal Practice**, v.6, p.655-670, 1990.

FRANÇA, P.M. **Use of ionophores in the feeding of Santa Inês lambs for meat production.** Thesis (doctorate). Lavras: UFLA, 2011. 123p.

FURTADO, R.N.; CARNEIRO, M.S.S.; CÂNDIDO, M.J.D. et al. Nitrogen balance and rumen evaluation in male and female sheep fed diets containing castor bean cake under different treatments. **Semina: Ciências Agràrias**, Londrina, v.35, n.6, p.3237-3248, 2014.

GARCIA, C.A. **Energy levels in *creep-feeding* lambs.** Botucatu: Universidade Estadual Paulista, 2002. 60p. Thesis (Doctorate in Zootechnics) - Paulista State University, 2002.

GERMAN, A.J., HALL, E.J., DAY, M.J. Chronic intestinal inflammation and intestinal disease in dogs. **Journal of Veterinary Internal Medicine**, v.17, n.1, p.8-20, 2003.

GOMES, F.H.T.; CÂNDIDO, M.J.D.; PEREIRA, E.S.; et al. Weight and yield of commercial cuts in sheep fed diets containing castor bean cake. In: Congresso Nacional de Zootecnia, 20, Palmas, 2010. **Anais...** Palmas: Zootec, 2010. CD - ROM.

GONZAGA NETO, S.; SOUZA, W.H.; VINAGRE, A.C.R.; et al. Biometric measurements and carcass yield of Santa Inês and Dorper x santa Inês crossbred sheep. In: Brazilian Congress of Animal Science, 2005. **Proceedings...** Campo Grande: ZOOTEC, 2005.

HALL, E.J. Clinical laboratory evaluation of small intestinal function. **Veterinary Clinics of North America - Small Animal Practice**, v.29, n.2, p.441-469, 1999.

HALL, M.B. Recent advances in non-fibrous carbohydrates in dairy cow nutrition. In: SIMPÓSIO INTERNACIONAL DE BOVINOCULTURA DE LEITE: Novos conceitos em nutriçâo, 2., 2001, Lavras. **Proceedings...** Lavras: Federal University of Lavras, 2001. p.149-159.

HOFFMAN, W. A.; PONS, J.A.; JANER, J.L. The sedimentation concentration method in *Schistossoma mansoni*. **Journal of Public Health Tropical Medicine**, v.9, p.281-298, 1934.

JOLEY, W.R. BARDSLEY, K.D. Ruminant coccidiosis. **Veterinary Clinics of North America - Food A**; v.22, p.613-621, 2006.

JOHNSON, A.D. Sample preparation and chemical analysis of vegetation. In: MANETJE, L.T. (Ed.) **Measurement of grassland vegetation and animal production**. Aberystwyth: Commonweath Agricultural Bureaux, 1978, p. 96-102.

KANEKO, J.J., Harvey J.W. & Bruss M.L. **Clinical biochemistry of domestic animals**. 6.ed. San Diego: Academic Press, 2008. 916p.

KUMAR, O.; SUGENDRAN, K.; VIJAYARAGHAVAN, R. Oxidative stress associated hepatic and renal toxicity induced by ricin in mice. **Toxicon**, v.41, p.333-338, 2003.

LAEMMLI, U.K. Cleavage of structural proteins during the assembly of head of bacteriophage T4. **Nature**, v.227, p.680-685, 1970.

LIMA, J. D. Coccidiosis of domestic ruminants. **Revista Brasileira de Parasitologia Veterinària**, v.13, supplement 1, p.9-13, 2004.

MANERA, D.B.; VOLTOLINI, T. V.; YAMAMOTO, S. M. et al. Productive performance of grazing sheep supplemented with concentrates containing fruit bran coproducts. **Semina: Ciências Agràrias**, Londrina, v.35, n.2, p.10131022, 2014.

MATOS, M.S.; MATOS, P.F. **Laboratòrio clinico médico-veterinàrio**. 2.ed. Rio de Janeiro: Atheneu, 1988. 238p.

McMANUS, J.F.; MOWRI, R.W. **Staining methods - histological and histochemical.** New York: Hoeber, 1960. 423p.

MENDES, C.Q.; SUSIN, I.; PIRES, A.V. Performance of lambs receiving starter concentrates with different crude protein contents. In: REUNIÂO ANUAL DA SOCIEDADE

BRAZILEIRA DE ZOOTECNIA, 40. 2003. **Anais...** Santa Maria, 2003. CD.

MIRZA, S.N.; PROVENZA, F.D. Preference of the mother affects selection and avoidance of foods by lambs differing in age. **Applied Animal Behavior Science**, v.28, p.255-263, 1990.

MITCHELL, E.S.E.; SMITH; R.P.; ELLIS-IVERSEN, J. Husbandry risk factors associated with subclinical coccidiosis in young cattle. **Veterinary Journal**, v.193, p.119-123, 2012.

MORAES, J.C.F.; SOUZA, C.J.H. de; GONÇALVES, P.B.D. Control of oestrus and ovulation in cattle and sheep. In: **Biotechniques applied to animal reproduction**. Sao Paulo: Varela, 2002, c.3, p.25-56.

MOREIRA, J.F.C.; RODRiGUEZ, N.M.; FERNANDES; P.C.C. et al. Protein concentrates for cattle. 1. *in situ* digestibility of dry matter and crude protein. **Arquivo Brasileiro de Medicina Veterinària e Zootecnia**, Belo Horizonte, v.55, n.3, p.315-323, 2003.

MORGADO, E.S.; EZEQUIEL, J.M.B.; GALZERANO, L. et al. Consumption, digestibility and nitrogen balance of lambs fed high starch or neutral detergent soluble fiber associated with sunflower oil. **Semina: Ciências Agràrias**, Londrina, v.35, n.1, p.457-466, 2014.

OLIVEIRA, G.A. **Lambs of three sex classes fed individually or in groups.** Thesis (Doctorate). Itapetinga-BA: UESB, 2014. 83f.

OLIVEIRA, A.S. **Co-products of castor and sunflower seed oil extraction in ruminant feed**. 2008. 166f. Thesis (Doctorate in Animal Production) - Federal University of Viçosa. Viçosa, MG, 2008.

OLIVEIRA, A.S.; OLIVEIRA, M.R.C.; CAMPOS, J.M.S. et al. In vitro ruminal degradation of ricin and its effect on microbial growth. **Animal Feed Science and Technology**, v.157, p.41-54, 2010.

PEREIRA, T.C.J. **Substitution of corn by algaroba bran in pelleted diets for lambs.** Thesis (Doctorate). Itapetinga-BA: UESB, 2012. 72f.

POLI, C.H.E.C.; MONTEIRO, A.L.G.; BARROS, C.S. et al. Beef sheep production in four production systems. **Revista Brasileira de Zootecnia**, v.37, n.4, p.666-673, 2008.

PORTO, M.O.; PAULINO, M.F.; VALADARES FILHO, S.C. et al. Energy sources in multiple supplements for Nelore calves in creep-feeding: productive performance, consumption and nutrient digestibility. **Revista Brasileira Zootecnia**, v.38, n.7, p.1329- 1339, 2009.

RADOSTITS, O.M., GAY, C.C., HINCHCLIFF, K.W. et al. **Veterinary medicine: a textbook of the diseases of cattle, horses, sheep, pigs and goats**. 10.ed. Edinburgh: W.B. Saunders, 2007. 2156p.

RIBEIRO, T.M.D.; MONTEIRO, A.L.G.; PRADO, O.R. et al. Animal performance and carcass characteristics of lambs in four production systems. **Revista Brasileira de Saùde e Produçâo Animal**, v.10, n.2, p.366- 378, 2009.

RODRIGUES, G.H.; SUSIN, I.; PIRES, A.V. et al. Substitution of corn by citrus pulp in high concentrate rations for confined lambs. **Ciência Rural,** Santa Maria, v.38, n.3, p.789-794, 2008.

SA, C.O.; SIQUEIRA, E.R.; SA, J.L. et al. Influence of photoperiod on feed intake, production and milk composition of Bergamâcia ewes. **Pesquisa Agropecuâria Brasileira**, v.40, n.6, p.601-608, 2005.

SANTOS, T.R. **Treated castor bean bran, dry and wet, in lamb feed.** Dissertation (Master's Degree in Animal Science) - Viçosa: UFV, 2011.39p.

SEEBERGER, P.H; WERZ, D.B. Chemical glycomics as basis for drug discovery. In: SCHREIBER, S.L.; KAPOOR, T.M.; WESS, G. (Eds.). **Chemical biology: from small molecules to systems biology and drug design**. Weinheim: Wiley-VCH, 2007. ch.11, p.668-691.

SEVERINO, L.S. **What we know about castor beans.** Campina Grande: Embrapa Cotton, Documentos, 134, 2005. 31p.

SVENSSON, C. Excretion of *Eimeria alabamensis* oocysts in grazing calves and young stock. **Journal of Veterinary Medicine B**, v.47, p.105-110, 2000.

SILVA, D.F. **Effect of prebiotics added to enteral diets on the translocation of *Klebsiella spp.* in mice**. 2004. 45f. Thesis (Master's Degree in Agricultural Microbiology) - Federal University of Viçosa, Viçosa, MG, 2004.

SILVA, T.D.F. ***Creep-feeding* for lambs with fed or pelleted rations**. Dissertation (Master's Degree in Zootechnics) - Maringà: UEM, 2009.

SILVA, D.J.; QUEIROZ, A.C. **Food analysis (chemical and biological methods)**. 3.ed.

Viçosa, MG: Editora UFV, 2002. 235p.

SOTOMAIOR, C. S.; ROSALINSKI-MORAES, F.; SOUZA, F. P.; MILCZEWSKI, V.; PASQUALIN, C. A. **Gastrointestinal parasites of sheep and goats - control alternatives**. Technical Information Series, n.80. Curitiba: EMATER Institute, 2009. 36p.

SOUSA, V.S. **Performance, carcass characteristics and blood parameters in Santa Inês sheep supplemented with *Crambe abyssinica*.** Thesis (Doctorate). Brasilia: University of Brasilia, 2014, 72p.

SNIFFEN, C.J.; O'CONNOR, J.D., VAN SOEST, P.J. et al. A net carbohydrate and protein system for evaluating cattle diets. 2. Carbohydrate and protein availability. **Journal of Animal Science**, v.70, n.11, p.3562-3577, 1992.

SUSIN, I. Sheep nutritional requirements and feeding strategies. In: SILVA SOBRINHO, A.G. (Org.). **Nutriçâo de Ovinos**. Jaboticabal: FUNEP, 1996. p.119141.

THORHALLSDOTTIR, A., PROVENZA, F., & BALPH, D. Ability of lambs to learn about novel foods while observing or participating with social models. **Applied Animal Behavior Science**, v.25, p.25-33, 1990.

URQUHART, G.M.; ARMOUR, J.; DUNCAN, J.L. et al. (Eds). **Veterinary parasitology.** 2.ed. Rio de Janeiro: Guanabara Koogan, 1998. p.196- 202.

USHIDA, K., LASSALAS, B., JOUANY, J.P. Determination of assay parameters for RNA analysis in bacterial and duodenal samples by spectrophotometry. Influence of sample treatment and preservation. **Reproduction Nutrition Development**, v.25, n.6, p.1037-1046, 1985.

VALADARES, R.F.D.; BRODERICK, G.A.; VALADARES FILHO, S.C. et al. Effect of replacing alfalfa silage with high moisture corn on ruminal protein synthesis estimated from excretion of total purine derivatives. **Journal of Dairy Science**, v.8, p.2686-2696, 1999.

VALADARES FILHO, S.C.; VALADARES, R.D.F. Recent advances in protein in dairy cow nutrition. In: SINLEITE - INTERNATIONAL SYMPOSIUM ON NEW CONCEPTS IN NUTRITION, 2, Lavras. **Proceedings...** Lavras, p.229-247, 2001.

VERBIC, J.; CHEN, X.B.; MACLEOD, N.A. et al. Excretion of purine derivatives by ruminants.Effect of microbial nucleic acid infusion on purine derivative excretion by steers.

Journal of Agricultural Science, v.114, p.243-248, 1990.

WEISS, W.P. Energy prediction equations for ruminant feeds. In: CORNELL NUTRITION CONFERENCE FOR FEED MANUFACTURERS, 61, 1999, **Proceedings...** Ithaca: Cornell University Press, p.176-185, 1999.

ZUNDT, M.; OLIVEIRA, K.P.; AMBIEL, A.C. et al. Performance of male and female dorper crossbred lambs in creep feeding receiving pelleted feed until weaning. **Colloquium Agrariae**, v.10, n.1, p.26-32, 2014.

MIX
Papier aus verantwortungsvollen Quellen
Paper from responsible sources
FSC® C105338

Printed by Books on Demand GmbH, Norderstedt / Germany